ALLYN AN

Mind Matters™

· CD-ROM FOR WINDOWS® AND MACINTOSH® ·

JAMES L. HILTON
UNIVERSITY OF MICHIGAN

CHARLES W. PERDUE
WEST VIRGINIA STATE COLLEGE

USER'S GUIDE

Allyn and Bacon
Boston · London · Toronto · Sydney · Tokyo · Singapore

For technical support of this product please call 1-800-677-6337

A Pearson Education Company
Needham Heights, Massachusetts 02494

Internet: www.abacon.com

ISBN 0-205-32179-8

Printed in the United States of America

10 9 8 7 6 5 4 3 01 00

TABLE OF CONTENTS

A unique learning tool, The Allyn and Bacon Mind Matters CD-ROM helps students explore psychology by combining interactivity with clear explanation, fostering active learning and reinforcing core concepts in the introductory psychology course. This CD-ROM contains a wide range of learning opportunities including activities with immediate scoring and feedback, video clips of historic experiments and current research, animations, simulations, and an interactive glossary of key terms. Introductions and conclusions place all activities in a rich context to enhance comprehension. Learning is reinforced through two forms of student assessment, Rapid Reviews after each topic and more extensive Quick Quizzes after each Unit. Easy navigation allows students to work through each Unit in a linear or non-linear format. The Allyn and Bacon Mind Matters CD-ROM is accompanied by an extensive Faculty Guide with descriptions of all activities, outlines, text correlation guides, and additional test questions for each Unit. Enjoy!

Welcome

As the authors of the **Allyn and Bacon Mind Matters** CD-ROM, we would like to take this opportunity to welcome you and tell you a little bit about the CD-ROM. Four years ago we were watching James Hilton's son play on the computer. At the time, Michael Hilton was 5 years old and, although he could not yet read, we were amazed at the fact that he had no trouble playing on the computer. As we watched, we also began to realize that he was "learning" an awful lot while playing. More importantly, he was learning with a joy and enthusiasm that is too rarely achieved in the classroom. From that moment, we were hooked. We became obsessed with the idea of finding ways to use digital technology to enhance the teaching and learning of psychology. We looked at every piece of software we could find. We bought all kinds of CD-ROMs, from digital cookbooks to computer games. We drove the people around us nuts. Fortunately, just as the patience of those we love began to wear thin, we convinced Allyn and Bacon to join us in the quest to use digital technology to make psychology more engaging, interactive, informative, and fun.

This CD-ROM is designed to be compatible with virtually any introductory psychology textbook. It includes units devoted to history, methods, biopsychology, learning, memory, sensation, and perception. Each unit, in turn, contains a series of self-contained modules that cover "core" psychological concepts through a combination of text, graphics, humor, and activities. Our goal is to present and integrate psychological concepts in ways that invite users to explore the "science of the mind."

James L. Hilton • Charles W. Perdue

HOW TO USE THIS PROGRAM

INTRODUCTION

Welcome to Allyn and Bacon Mind Matters. Below you will find a full tour on how to use this product. This tour will guide you through the easy-to-use navigation features of this unique CD-ROM. The tour is divided into three parts:

- General Navigation
- The Table of Contents
- The Glossary

For Help while you are working within the program, you can also click on the *How to use this CD-ROM* icon on the Main Menu.

All sections of this CD-ROM can be accessed from the Main Menu. There are five units and these units are further divided into topics. Within the topics there are sections that contain the content and special features. Below is a brief description of some of these features.

- Pop-Up windows give you a more in-depth look at a particular subject. These are identified within the program by a star next to the title.

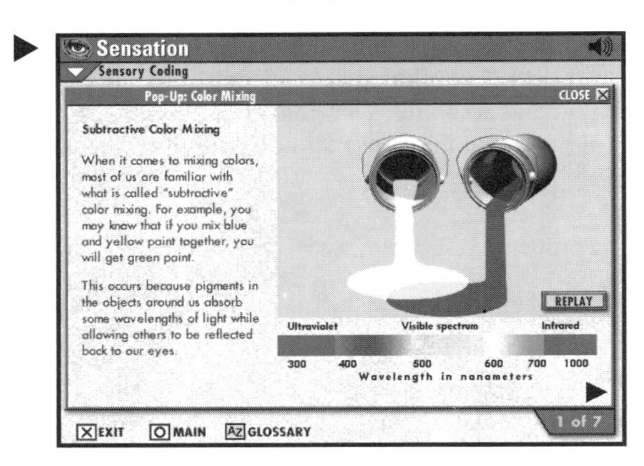

- A wide variety of activities and simulations to help you master the concepts.
- Videos of both historical footage and contemporary topics.
- Animations which illustrate many of the concepts.

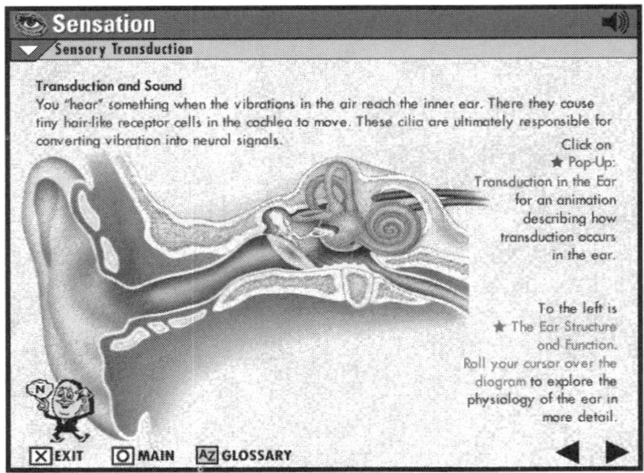

• Nugget Man offers a fun fact.

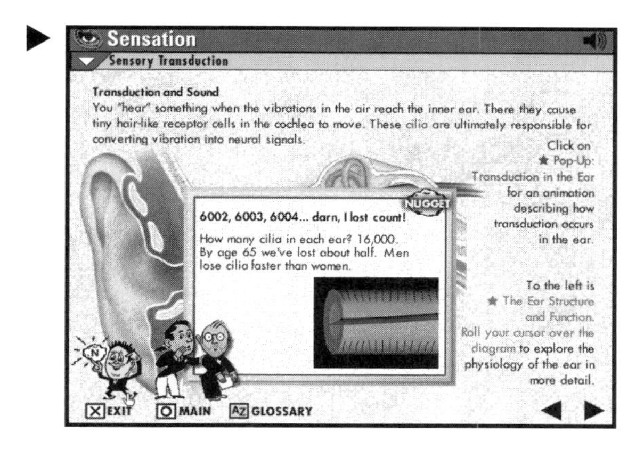

• Rapid Reviews are short practice tests to help you review the topic you just finished. ▶

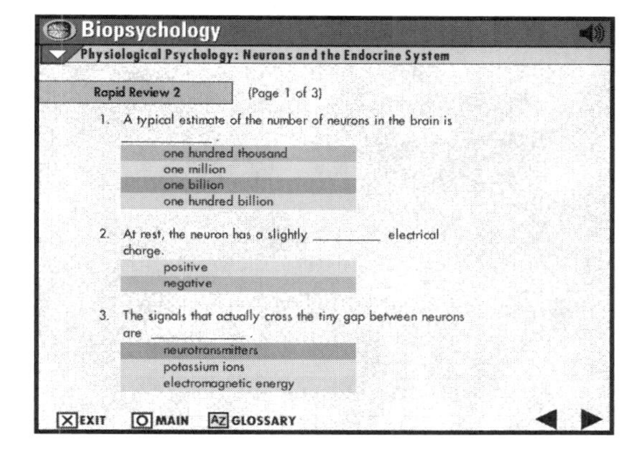

• Quick Quizzes are an-end-of unit test with scoring and references to the material you need to review. ▶

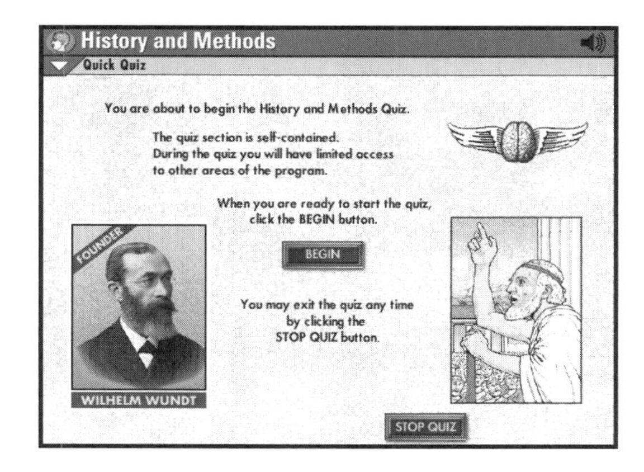

• A full interactive glossary can be viewed by unit

GENERAL NAVIGATION

1. You can access virtually the entire CD-ROM from the main menu. Start on the left-hand side by moving the mouse to one of the unit titles.

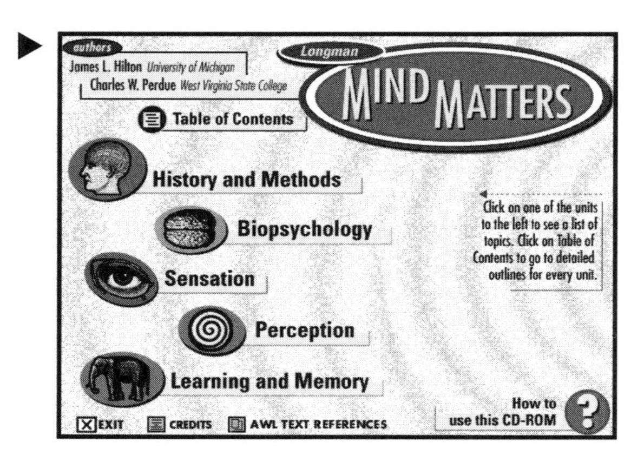

2. For example: Click on "Sensation" and follow the dotted line that appears to the right. The dotted line will lead you to a list of the topics within the unit
3. Click on the word "Introduction" to read the unit from beginning to end in a linear fashion.
4. Click on any other topic below "Introduction" to move immediately to that section.
5. To advance the story, click on the right-pointing triangle at the bottom right of the screen.
6. The left pointing triangle will take you back to the previous screen.

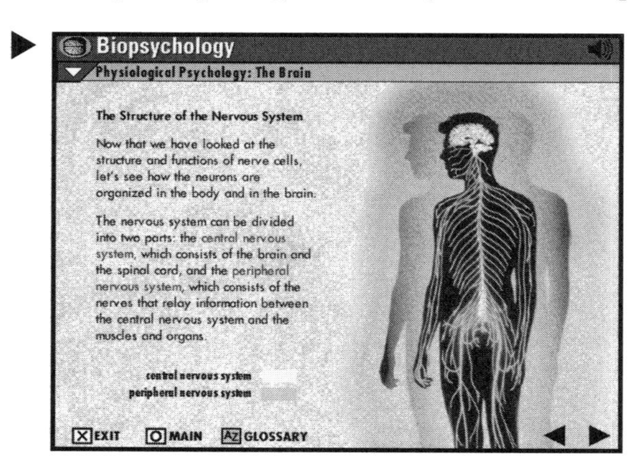

7. In the story there is a title bar at the top of the screen. The upper portion indicates what unit you are in. The lower portion indicates the current topic of that unit that you are covering.

8. Click on the left side of the title bar to activate a "drop down" menu.
9. Move the cursor to the topic of the unit you wish to explore.
10. Move to the second drop-down menu and click on a heading to move to a section within that topic, for greater specificity.

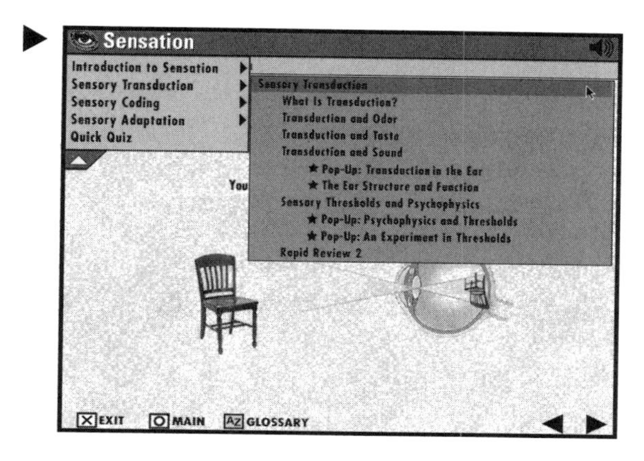

11. Glossary words are highlighted blue. Click on them to move to the glossary and read the definition of the psychology term. Click on the "X" at the top right to return to the original topic.
12. Move the cursor to the Nugget Man to read a fun fact or information nugget about the topic at hand.
13. Move the cursor to the footnote icon to read the footnote.
14. The blue text with the star icon indicates a Pop-Up window. Click on the blue text to explore an activity, a video, or an in-depth look at an area of psychology.
15. There are a number of movies. Each movie has a play button and a rewind button below it. The play button becomes a pause button once the movie has been started. Simple animations play automatically.
16. To return to the story from a movie, click on Close.
17. Click on the sound button located at the upper right-hand corner of the screen to toggle the sound on or off.
18. Red-highlighted text in the main story gives instructions for an activity on the current screen.

19. The Glossary button at the bottom of every page takes you to the glossary.
20. The Exit button at the bottom of every page allows you to exit Allyn and Bacon Mind Matters.
21. The Main button at the bottom of every screen takes you back to the Main Menu.

THE TABLE OF CONTENTS

T1 You can go directly to any section of this CD-ROM by clicking on the Table of Contents button located on the Main Menu.

T2 All five units are listed on a Table of Contents bar at the top of the screen.

T3 Click on a unit title to "drop down" a list of the topics within the unit.

T4 Move the mouse to a topic in that list to automatically "drop down" a second list which displays the headings within that topic.

T5 To go directly to the first screen of a topic, click on the topic title.

T6 Move the cursor to the second list and click on a heading to go to the beginning of that section within the topic.

T7 Starred headings indicate an activity or a Pop-Up window.

THE GLOSSARY

G1 You can click the Glossary button at the bottom of the page to move to the glossary from any screen.

G2 You can also click on blue-highlighted glossary words within the main story to take you to specific words within the glossary.

G3 The definition of the glossary word you have chosen appears in the frame to the left of your screen.

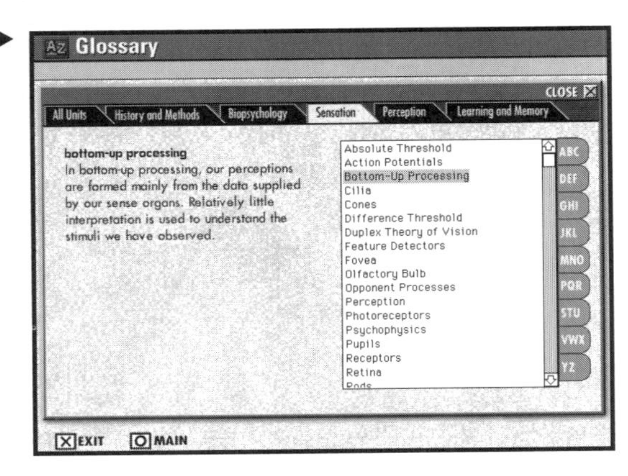

G4 Click on a unit's tab, located at the top of the glossary window, to see a list of all the glossary words in that unit.

G5 The list of glossary words from that unit will appear in the scrolling window on the right-hand side of the screen.

G6 Click on the letter buttons to the right of the glossary list to quickly scroll to the chosen letter.

G7 Click on the desired glossary word to display its definition.

G8 Click on the All Units tab located at the top of the glossary window to see a complete list of all glossary terms in this CD-ROM.

TABLE OF CONTENTS

SYSTEM REQUIREMENTS

MACINTOSH INSTALLATION REQUIRES

- 68040/66 processor; System 7, or greater
- 16 MB physical RAM free
- 8 MB free hard disk space
- 256 color monitor (video quality will improve when monitors are set to thousands of colors)
- Double-speed CD-ROM drive (4x recommended)
- QuickTime installation (included)

WINDOWS INSTALLATION REQUIRES

- 486/66 processor or Pentium, Windows 3.1 (Windows 95 or Windows 98 recommended), or Win NT
- 16 MB physical RAM free
- 8 MB free hard drive space
- SVGA card for 256 color display, 640 x 480 (video quality will improve when monitors are set to thousands of colors)
- Double-speed CD-ROM drive (4x recommended)
- 8-bit SoundBlaster or 100% compatible sound card (16-bit recommended) speakers or headset
- QuickTime 2.12 installation (included)

INSTALLATION

ON A MACINTOSH COMPUTER

- Close any open applications or windows
- Turn off Virtual Memory in the Memory Control Panel
- Place the disc into your CD-ROM drive
- A window titled Allyn and Bacon Mind Matters will appear on the desktop
- Double-click on Allyn and Bacon Mind Matters Setup
- A Select Destination Folder box will appear
- Choose your destination folder and press the Extract button
- A folder titled "Allyn and Bacon Mind Matters" will appear in your chosen location
- Open this folder and click on the application titled Allyn and Bacon Mind Matters

ON A WINDOWS 95 COMPUTER

- Close any open applications or windows
- Place the disc into your CD-ROM drive
- Double-click on My Computer icon
- Double-click on the CD-ROM drive icon, Awlmind
- Double-click on install.exe
- Follow on-screen instructions
- You will automatically be prompted to install QuickTime, a necessary helper program, this product requires the installation of QuickTime 2.1.2

INSTALLATION INSTRUCTIONS FOR QUICKTIME 2.1.2

- Read the software license
- On the window "Begin Install"—click Install
- On the window "Check Existing Versions"—click Start
- On the window "Complete Install"—click Install
- On the window "Success"—click Exit and then confirm by clicking Yes
- The AB Mind Matters icon will appear in a window.
- Double-click on the AB Mind Matters icon or click on the Start button and Select AB Mind Matters from the program menu.

ON A WINDOWS 3.1 COMPUTER

- Close any open applications or windows
- Insert the Allyn and Bacon Mind Matters CD-ROM into your disk drive
- In the Program Manager, open File Manager
- Click the drive icon for your CD-ROM drive
- Open the win3_1 folder on the AB Mind Matters CD
- Double-click the instl3_1.exe application
- Follow the onscreen instructions
- You will automatically be prompted to install QuickTime, a necessary helper program, this product requires the installation of QuickTime 2.1.2
- Follow the onscreen instructions
- When the installation is finished, the AB Mind Matters icon will appear in a window.
- Double-click on the AB Mind Matters.

NOTES

NOTES

LICENSING AGREEMENT

You should carefully read the following terms and conditions before opening this disk package.
Opening this disk package indicates your acceptance of these terms and conditions.
If you do not agree with them, you should promptly return the package unopened.

Allyn and Bacon provides this Program and License its use. You assume responsibility for the selection of the Program to achieve your intended results, and for the installation, use, and results obtained from the Program. This License extends only to use of the Program in the United States or countries in which the Program is marketed by duly authorized distributors.

License Grant
You hereby accept a nonexclusive, nontransferable, permanent License to install and use the Program on a single computer at any given time. You may copy the Program solely for backup or archival purposes in support of your use of the Program on the single computer. You may **not** modify, translate, disassemble, decompile, or reverse engineer the Program, in whole or in part.

Term
This License is effective until terminated. Allyn and Bacon reserves the right to terminate this License automatically if any provision of the License is violated. You may terminate the License at any time. To terminate this License, you must return the Program, including documentation, along with a written warranty stating that all copies of the Program in your possession have been returned or destroyed.

Limited Warranty
The Program is provided "As Is" without warranty of any kind, either express or implied, including, but **not** limited to, the implied warranties or merchantability and fitness for a particular purpose. The entire risk as to the quality and performance of the Program is with you. Should the Program prove defective, you (and **not** Allyn and Bacon or any authorized distributor) assume the entire cost of all necessary servicing, repair, or correction. No oral or written information or advice given by Allyn and Bacon, its dealers, distributors, or agents shall create a warranty or increase the scope of its warranty.

Some states do **not** allow the exclusion of implied warranty, so the above exclusion may **not** apply to you. This warranty gives you specific legal rights and you may also have other rights that vary from state to state. Allyn and Bacon does **not** warrant that the functions contained in the Program will meet your requirements or that the operation of the Program will be uninterrupted or error free.

However, Allyn and Bacon warrants the disk(s) on which the Program is furnished to be free from defects in material and workmanship under normal use for a period of ninety (90) days form the date of delivery to you as evidenced by a copy of your receipt.

The Program should **not** be relied on as the sole basis to solve a problem whose incorrect solution could result in injury to a person or property. If the Program is employed in such a manner, its is at the user's own risk and Allyn and Bacon explicitly disclaims all liability for such misuse.

Limitation of Remedies
Allyn and Bacon's entire liability and your exclusive remedy shall be:

1. The replacement of any disk not meeting Allyn and Bacon's "Limited Warranty" and that is returned to Allyn and Bacon or
2. If Allyn and Bacon is unable to deliver a replacement disk or cassette that is free of defects in materials or workmanship, you may terminate this Agreement by returning the Program.

In no event will Allyn and Bacon be liable to you for any damages, including any lost profits, lost savings, or other incidental or consequential damages arising out of the use or inability to use such Program even if Allyn and Bacon or an authorized distributor has been advised of the possibility of such damages of for any claim by any other party.

Some states do **not** allow the limitation or exclusion of liability for incidental or consequential damages, so the above limitation or exclusion may **not** apply to you.

General
You may **not** sublicense, assign, or transfer the License of the Program. Any attempt to sublicense, assign, or transfer any of the rights, duties, or obligations hereunder is void.

This Agreement will be governed by the laws of the State of Massachusetts.

Should you have any questions concerning this Agreement, or any questions concerning technical support, you may contact Allyn and Bacon by writing to:

Allyn and Bacon
A Pearson Education Company
160 Gould Street
Needham Heights, MA 02494

You acknowledge that you have read this Agreement, understand it, and agree to be bound by its terms and conditions. You further agree that it is the complete and exclusive statement of the Agreement between us that supersedes any proposal or prior Agreement, oral or written, and any other communications between us relating to the subject matter of this Agreement.

Notice To Government End Users
The Program is provided with restricted rights. Use, duplication, or disclosure by the Government is subject to restrictions set forth in subdivison (b)(3)(iii) of The Rights in Technical Data and Computer Software Clause 252.227-7013.